To JACOB,
YAy SCIENCE!

Dustin J. Wilgers

Illustrated by Nate Knott

Savanna Spider, Super Scientist, Goes to School:
Science Saves the Day!
By Dustin J. Wilgers

Illustration by Nate Knott
Editing and book design by Jan Gilbert Hurst

Published by Author's Voice
1315 East Euclid, McPherson, KS 67460
www.AuthorsVoicePublishing.com

ISBN: 978-0-9970062-8-5
LCCN: 2018940686

Printed and bound in the United States of America by Mennonite Press, Inc., Newton, Kansas.

Dedications

From Dustin—

To my favorite field assistants, Noah and Hannah: Thank you for making my life story so exciting, funny, unpredictable, and totally unforgettable. I love you both. I can't imagine my plot without you.

To Autumn, my partner in this grand story called life: Thank you for supporting me regardless of how much the organisms that I am fascinated with creep you out. I love you and couldn't have done this without you.

From Nate—

To my spouse Mindy and our three boys, Kalen, Edison, and Jace: Thanks for giving my life so much love and laughter.

To all the great teachers I've had along the way, Gary "Captain Clay" Martin, Kathy Stastny, and Mary Ann Anderberg: Thanks for stoking the fire!

Acknowledgments

Thanks to the Kansas Department of Wildlife, Parks, and Tourism for funding in part this publication through their Chickadee Checkoff of the Nongame Wildlife Improvement Program. All persons filing a Kansas income-tax form have the opportunity to make a special contribution, through the convenience of the tax form or direct donations sent to Chickadee Checkoff, 512 SE 25th Ave., Pratt, KS 67124. These contributions are earmarked for conservation of nongame wildlife. Do something wild! Make your mark on the tax form for nongame wildlife.

A portion of the sales of this book will be donated to the Chickadee Checkoff Nongame Wildlife Improvement Program.

Chapters

This story personifies the arthropod characters and makes them behave in some ways that they can't and don't behave naturally. However, there are several examples of real biology behind the story and some of the characters' behaviors. Look for the 🕸 symbols throughout the story along with the "Science behind the Story" section at the back of the book that highlight some of these interesting science facts.

1

First Day of School

BUZZZZZZZ!!!!! Savanna Spider jumped out of her web hammock so quickly that she bumped into the microscope beside her bed and knocked over a bowl that held this crazy-weird bug she found in her backyard yesterday. She rubbed all eight of her eyes to see if she could find it crawling across the floor, but it was too late. She would have to wait to find her specimen and figure out exactly what it was another time, because today was a big day.

"Hurry downstairs for breakfast. I made your favorite, a fabulous fly smoothie!" her mom shouted from the kitchen.

Because spiders lack teeth, they consume their food, which is sometimes larger than they are, by liquefying it outside their mouths with their stomach acid and then sucking up the juices. They are essentially drinking a "prey smoothie" when they eat, just like Savanna did for breakfast.

Sitting at the table sipping her smoothie, Savanna tapped her feet, all eight of them, to her favorite song. Before long, her dad joined in across the table, singing, ". . . out came the sun and dried up all the rain." The whole family joined together for their favorite part, "Then the itsy bitsy spider climbed up the spout agaaaaiiiiiiiiiiiiiinnnnnnnnn."

"Savanna," her dad said with a smile, "I can't believe my cute little spiderling is old enough to be in school! Let's try to avoid doing any experiments on your first day, and don't start anything on fire. Okay, my little scientist?"

Savanna nodded.

"Some days I'm surprised our house is still standing after this summer." Her dad said, grinning at her. "Most importantly though, make sure to be good, have fun, and learn a lot! Now hurry upstairs and put your shoes on so you won't be late."

Savanna scrambled to her closet to find her favorite outfit, her lab coat, and her favorite four pairs of shoes. To her surprise, all of her shoelaces were gone.

"MOM!!!!" Savanna screamed, "All of my silk shoelaces are missing from my shoes!"

She ran across the hall to find her older brother, Woody. "Do you know what happened to my shoelaces?"

"Sorry, Savanna. We were out of flies last night, and I was really hungry for a snack."

"Ugghh… You can't be serious! Now what am I going to wear to my first day of school!" Savanna cried.

"Now Woody," their mother said calmly, "I know how your father and I encourage recycling, but that doesn't mean eating your sister's shoelaces. Don't worry Savanna, I can spin you some new laces in no time." 🕸

As soon as Savanna's laces were made and tied, she got up and raced out the door.

🕸Spider silk is both really strong and stretchy. Pound for pound it is stronger than steel. It would make great shoelaces. Not all silk is sticky; spiders can spin different types of silk based on what the silk needs to do. Spiders do eat their own silk. Many web-building spiders who eat their own silk recycle that material into the webs they build in the coming nights.

"Not so fast, Miss. Show me what's in your lab coat pockets!" Mom yelled. Savanna rolled her eyes, turned around, and dumped out the contents of her pockets: a magnifying glass, a collection jar full of slime mold, a pair of tweezers, a pad of paper, a pen, and a pair of beetle grubs. 🕸

🕸Scientists try to understand the world around them through careful observation. To do this, scientists often use tools to collect things and see details up close that they can't normally see. Sometimes a pen and paper are their most important tools, so they can record what they see to study later.

"Let's leave the living things at home today; not everyone is as fond of slime molds and grubs as you are," Mom explained.

* * *

By the time Savanna finally got out the door, she looked down the street to see her bus driving away.

"Oh no, now I'll never get to school on time!"

The thought of being late to her first day of school sent shivers down the hairs of her legs. Her school was ten blocks away. She had to hurry.

Just as she began to run down the street toward her school, a big gust of wind came from behind and almost pushed her over. "Of course!" Savanna thought, "My only chance of getting to school on time is by flying."

Savanna let out a long line of silk into the wind. In no time she was picked up off the ground like a balloon and headed toward her school.

"Up, up, and away!" She giggled. Savannah was so excited that she didn't realize how fast she was going. She should have remembered what her dad taught her as a little spiderling. Flying with the wind is fast and fun, but it is hard to control. Before she knew it, Savanna was past the school and needed to backtrack. 🕸

🕸Some spiders can use silk to fly. This is a common way that some young spiderlings disperse to new environments once they hatch from the egg sac. Like Savanna, they are at the mercy of the wind and have little control on the distance they travel. Spiders can parachute as little as a few yards or even up to hundreds of miles before settling down to land.

2

Arthur O. Pod Elementary School

By the time Savanna straggled into Arthur O. Pod Elementary School, she was fifteen minutes late, and no other students were in sight. As she turned to head down the hallway toward the office, Savanna ran straight into the chest of Principal Honeybee.

"Whoa, excuse me, but what are you doing in the halls? Shouldn't you be in your classroom?" Principal Honeybee asked.

Savanna blurted out every explanation she could think of: "I'm sorry ma'am; my brother ate my shoelaces; I missed my bus; and then the wind…"

"Okay, okay," Principal Honeybee interrupted. "Do you know where your classroom is?"

"Uh… No. Can you help me? Do you know where I need to go?" Savanna asked nervously, thinking she was already in trouble on the first day. This would not be a good report to her parents.

"I am the Queen of this school. I know who you are, Miss Spider, and I know exactly where your classroom is. I could tell you, but I think you may learn more from this if I let you figure out the answer for yourself. I notice you have a lab coat on. Are you ready for your first science assignment?" Principal Honeybee asked.

Savanna smiled, happy that the principal noticed her outfit. "Yes! I love science. It is my favorite subject!" Savanna proudly showed off all her tools, one by one.

"Very nice!" Principal Honeybee was impressed. "To find your classroom, you may have to use some of the tools in your pocket. I'm going to let you explore the school and visit each of the classes we have here. Careful observation is an important scientific skill.

Arthur O. Pod Elementary School has four classrooms: Class Diplopoda, Class Malacostraca, Class Insecta, and Class Arachnida. Instead of being divided by age, each class is organized based on other similarities the students share. Your task is to find the classroom that has the students that you share the most in common with. That is the class that you belong in."

3

Class Diplopoda
(dip-lo-po-da)

Savanna walked down the hall, excited. "I haven't even been to class yet, and I have my first science project!" When she came to the first classroom, Class Diplopoda, she pulled out her pen and paper, ready to take notes.

Looking through the window into the classroom, Savanna was so surprised that her mouthparts hit the floor. There were legs everywhere, hundreds of them!

She was in the middle of counting legs: 45, 46, 47... when out of nowhere, she heard "Ummm... hello, my name is Milton Millipede, wha-what are you doing, um,

looking into my classroom?" Milton muttered, hesitantly tapping Savanna on the back leg.

"Holy honeydew!" Savannah screamed as she jumped forward. "Where did you come from? You made me lose count. I'm trying to figure out if I belong in this class. What are you doing in the hallway?" Savannah asked Milton.

"I'm in Class Diplopoda, and they elected me hall monitor this year," Milton said, fumbling his badge.

"If you are in this class, may I ask you a couple of questions?" Savanna asked.

"Um, I don't know. We aren't supposed to be in the hallway," Milton said, pushing up his glasses.

When he did, Savanna noticed the antennae on the top of his head. "Hmm… interesting," she thought, "I don't have any of those."

Just then a voice came over the loud speaker, startling Savanna and Milton. "Good morning, I am Principal Honeybee and I want to welcome you all to a new school year. Our lunch staff is working hard to make plenty of food that each of our different students will like. If you are a predator and you are hungry, let your teacher know and please don't eat your classmates. That is all."

Savanna looked around. Milton was gone. "Milton, where did you go?" Savanna whispered.

"I'm down here," Milton answered, sounding kind of muffled. Savanna looked down and found Milton curled in a tight spiral.

"What are you doing down there? Get up!" Savanna said as she kicked him. "OUCH!" Savanna cried, "What are you made of, steel? I think I broke my tarsal claw."

"That's my hard, outer shell. Whenever I get scared, I curl up in a spiral and put all of my legs and underside in the middle to protect them," Milton explained. 🕸

When Milton was coiled in his spiral, all of his segments were very easy to see. And he had a lot of them, way more than she had.

"Really Milton, you can get up now; it's safe. I won't eat you. Speaking of legs, how many do you have?" Savanna asked.

🕸All spiders are predators, which means they must eat other animals. This likely includes their own schoolmates. No wonder Milton was nervous and jumpy! Millipedes, like Milton, often curl into a tight spiral to protect themselves from danger, using their hard exoskeleton.

"I normally have 60 legs, two pairs per segment, but I lost one leg last week when it got caught in the door, so 59 now. But, the next time I molt it will grow back," said Milton. ✦

"Does everyone in your class have that many legs?" Savanna asked, as she counted all eight of her own legs.

"Well, um, some have less and some have a lot more, but all of us in the class have two pairs on each segment," Milton explained.

"Whoa, that is a lot of legs. I bet your class is really good at soccer," Savanna said.

Milton finally smiled. "Yeah, we are the best team in school," Milton glowed, flexing ten of his short legs in the air. "Every once in a while, Class Malacostraca

---

✦Because arthropods have a hard exoskeleton, in order to grow, they must get rid of it and regenerate a new and larger one. When this happens, many arthropods can regenerate any legs they have lost since their last molt. Sometimes this takes a couple of molting cycles to produce a full-size leg.

will beat us, but I think they cheat; they get a little grabby with their pinchers."

Savanna was starting to figure things out. "Okay, you have a hard, outer shell like me, but you have antennae, your bodies are really segmented, and you have way more legs than I do. I don't think I belong in this Class. Maybe I should go check out Class Malacostraca. Where is their classroom?"

"Take a left at the next hallway. Their door is the first on the right. You'd better get there quickly, because you don't have a hall pass, and if Principal Honeybee catches you in the halls without one, both you and I will be in trouble," Milton warned, nervously drooping his antennae.

4

Class Malacostraca
(mal-uh-kos-truh-kuh)

For Milton's sake, Savanna quickly found the door for Class Malacostraca and quietly eased into the room. No one was in the room, but that wasn't the only odd thing about it. The room felt like she was sitting in a cloud. She knew how that felt after accidentally flying through one on the way to school this morning. Half of the desks in the classroom were underwater, and ceiling sprinklers sprayed mist over the other half of the room.

After about a minute imagining what kind of student could sit in an underwater desk, Savanna saw the door open. The teacher came in carrying something big on his back.

"Hello, may I help you?" he said in a deep voice. "Sorry, we were at the school pool for gym class. My name is Herman Hermit Crab. I'm the teacher of this class."

"Hmm…what is that on your back?" asked Savanna, staring with confusion.

"Oh, this is my shell. It's like my desk and home in one, but I get to carry it around with me. It's kind of convenient as a teacher to have everything you need with you all the time. Please excuse me just a second," Mr. Hermit Crab said as he went inside his shell and then quickly popped back out with his whistle.

TWEEEEET!!! "Okay students, please go to your seats. Francis, you are the only one with a strong enough claw; please go over to the aquarium door and let the rest of the class come in and go to their desks."

The students who came in through the door were bizarre! Some entered walking sideways, holding pinchers high in the air, while others carried shells on their backs like Mr. Hermit Crab. Francis came in last, waving a single, outrageously large claw.

"Excuse me, strong man coming through," Francis announced as he barged in and headed over to open an underwater door that led outside. Then much to Savanna's surprise, in swam the rest of the class, backwards!

As Francis shut the door, he closed it too quickly and caught the last student's long antenna in the door. "Hey!!!" she screamed. "Lift the door!"

"Sorry," Francis said, "I always forget your lobster antennae are so long because mine are short."

"That's okay, it didn't hurt. It helps to have a hard exoskeleton; that way clumsy crabs like you can't hurt me," she joked.

Once everyone had taken a seat, the only open desk out of the water was near Francis. Savanna sat down there.

"Hi, I'm Francis Fiddler Crab. What's your name?"

Francis said, extending the much smaller of the two claws. "I would shake with the big one, but most people get scared, and I'd also probably knock a whole bunch of things over."

"I'm Savanna. Okay, I have to know. What's with all the water?" she asked Francis.

Francis looked confused. "That's how we breathe. All of the students here have gills that only work with water. Even those of us who live on land need to stay wet."

Savanna had figured this classroom out quickly. "Ooooohhh, okay. I don't think I belong here, either. I don't have gills; I can't live in water; and I don't have antennae." Savannna was beginning to get antsy. "I'm running out of time, I need to find the class that I belong to fast," Savanna said shifting out of her seat toward the door.

"Maybe you're an insect, they don't have gills. Go down the hall. Third door on your left is Class Insecta. Good luck!" As Francis pointed in the direction of the classroom with his large claw, he knocked over the desk in front of him, which made such a loud noise that Mr. Hermit Crab disappeared into his shell as Savanna bolted out the door.

5

Class Insecta
(In-sek-ta)

Savanna could feel the floor shaking as she got close. The vibrations sounded like an army marching in unison around the classroom. As she peeked into the room for Class Insecta, she barely had enough time to get out of the way before the door slammed open. The students filed out behind the teacher, marching in a long straight line down the hallway.

Savanna saw three identical students marching in perfect unison at the front.

> Spiders are really good at feeling ground vibrations. They feel vibrations using organs in their legs called slit sensilla. In fact, some spiders communicate by making the ground vibrate with their pedipalps, legs, and/or abdomen.

"Hi, my name is Savanna," she said. "Where are you going?"

"Hi, I'm Sandy."

"I'm Mandy."

"I'm Candy."

"We're going to recess," they answered in unison. "Wanna come? Sorry, we can't wait, we have to stay in line; that is what ants do." As they marched past,

Savanna noticed the three ants each had three body parts: a head, a middle where the legs came from, and an abdomen at the end.

Behind the ants in line were students of all different shapes. Some had really large back legs that looked great for jumping, some had really small legs, and others had long front legs that were folded in front like they were praying. Toward the back of the line she saw some students flying with wings.

At the very end of the line, Savanna saw one student walking backwards while doing a handstand and rolling a soccer ball with its back legs. This was the weirdest thing that Savanna had seen all day. Really curious, Savanna asked, "What in the animal kingdom are you doing?"

"Whoa, you surprised me! I guess that is easy to do since I'm walking backwards," replied the student.

"And upside down," Savanna said, chuckling and leaning over as far as she could to get a better look at her new friend.

"My name is Daisy Dung Beetle. We are going to practice soccer during recess today. Our goal is to beat Class Diplopoda this year. I am the best in the class at dribbling, since I kind of do it for a living. 🕸 The class put me in charge of getting the ball to the playground," Daisy said proudly, finally standing upright and peeking over the ball.

"That sounds like fun! I'm trying to find the class I belong to. I hope it's this one. Can I come play soccer with your class at recess?" Savanna asked.

🕸Dung beetles are amazing insects that collect manure, form it into a ball, and then roll it across the field with their back legs, all the way to their burrow. Their young feed on these, using the nutrients in the dung to grow.

"Sure, your extra pair of legs will be a big help," Daisy observed.

"What do you mean?" Savanna said, confused.

"Well, you have eight legs and everyone in our

class has only six. Also, where are your wings? Mine are tucked up under this shell. No wings may be a problem. Most of the plays we are working on today are above the ground, because that is our advantage." Daisy explained as they headed out to the playground.

Walking onto the soccer field, Daisy announced to her class, "This is Savanna; she is going to play with us today. I think she will make a good goalie with all her legs." Everyone welcomed her to the field, especially Darnell Dragonfly, the goalie.

Darnell ran up to Savanna, saying, "Thank goodness you are here! My wings help me fly, but they are so delicate that the soccer ball has cracked them in a couple of places. Now they are so broken that I can't fly well enough to play in the field, which is why they keep on sending me to be goalie. No one else wants to do it."

"Hey, I have an idea," Savanna said. "I bet my sticky silk may be able to fix your wings. Let me get my magnifying glass out to take a closer look." As Savanna carefully added tiny sticky silk threads like stitches across the cracks in his wings, 🕸 she noticed that the wing was inserted into Darnell's middle body region. As she looked around, Darnell, the ants, and all the other students from Class Insecta on the field had the same three body regions. This was different from her body. She had only two.

In no time, Darnell's wing was mended and he was flying up to the front line to try and score a goal. "Thanks Savanna, you're the greatest!" Darnell squealed with joy. Savanna spent the rest of recess having a great time with her new friends on the soccer field.

🕸Some spider silk is sticky, which comes from tiny glue drop-lets added to the silk as it is formed leaving the spinnerets located on the spider's abdomen. This glue has to be really sticky to be able to hold insects to the web that are trying to fly away.

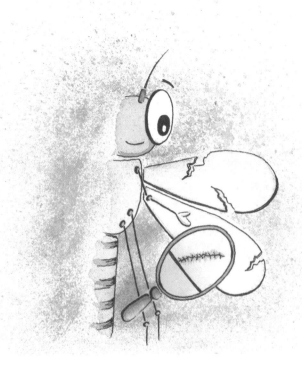

As the bell rang for the end of recess, Savanna lined up behind the three ant sisters thinking, "Three body parts, six legs, wings. Hmm... not quite like me." She knew now that even though she had tons of fun with a bunch of new friends, she could not return to the classroom with Class Insecta.

6

Class Arachnida

(ah-rak-ni-duh)

As Class Insecta filed into the building after recess, Savanna noticed the only classroom she hadn't visited yet: Class Arachnida. She knocked on the door and went inside.

"Ahhh... Ms. Spider, so glad you could join us. I'm Mr. Vinnie Garoon," she heard from the teacher standing at the front of the class.

"Wait, you know who I am?" Savanna said, as the hairs on her abdomen stood up.

"Sure, I've been wondering why you were late to class on your first day," he said.

"It's a long story. Are you saying this is my class, Arachnida?" Savanna said, rubbing her eyes with her pedipalps as she scanned the classroom.

"It sure is. Go have a seat by Sunny Solifugid. Now class, are you ready to play a little game?" asked Mr. Garoon. "Everybody stand up. Now I'm going to say a characteristic of our bodies, and if you don't have it, sit down." Everyone in the class loved games, so they quickly stood up. "If you don't have two body regions, sit down," said Mr. Garoon. No one sat down. "If you don't have eight legs, sit down," said Mr. Garoon. Savanna looked around and no one was sitting. In all the other classes she had visited during the day, either question would have eliminated them all.

Savanna raised a "leg" so the teacher would call on her. "So those are the main characteristics that we have in common in Class Arachnida?" she asked.

The teacher nodded. "Yep. Even though we all look pretty different, we share those things in common as arachnids." Savanna recognized that she shared each of those things in common with all of her classmates. Her search was over! "I get it! THIS IS MY CLASS!" Savanna screamed.

"Now that we have all that figured out, I think it's time to eat. Who is ready for some fly smoothies? The lunch ladies made them extra special for us today," Mr. Garoon announced.

The rest of the day went quickly for Savanna. She had such a good time with all of her classmates that she didn't want the day to end. The bell rang. As she walked out the doors, Principal Honeybee was waiting outside. "So, did you find your class?" she asked.

"Yes, I did. I had the most amazing day. From observation, I learned so much about all the students at this school. The students here have so much in common, but are so different, too. May I visit all the classrooms again tomorrow?"

"I'm glad you had so much fun, but let's not make this a habit, little miss scientist. There are so many other subjects that are fun and important too. Have a great night; see you on time in the morning," Principal Honeybee said with a smile.

Savanna hurried down to the bus, so she wouldn't miss it this time. As the bus drove away from the school, Savanna excitedly bounced on her seat. She couldn't wait to get home and tell her parents all about her crazy and wonderful first day of school and how science had saved the day. Yay science!

More Science behind the Story

Scientists use a system based on similarities to categorize, classify, and identify living organisms on our planet. They start with the broadest and most inclusive categories, becoming more specific each level below. The major scientific levels of classification are Domain, Kingdom, Phylum, Class, Order, Family, Genus, and Species.

The organization of Arthur O. Pod Elementary School follows the same general organization that science uses. The largest Phylum in the world is Phylum Arthropoda (organisms in this category are called Arthropods), all animals in this group are in Kingdom Animalia and Domain Eukarya. The animals in this group may look very different, but all of them share a few very-important characteristics with one another, including no backbone, a hard exoskeleton, and jointed legs and appendages.

Students in the school are organized into classrooms or "Classes," the next level below Phylum, based on the characteristics they share in common with one another and not students from other classes. The classes in this book are real Classes of arthropods and are just a small collection of the different arthropods found on our planet. Here are the Classes that Savanna visited throughout the day:

Class Diplopoda (Pronounced dip-lo-po-duh)

This group contains the millipedes. Members of this group have:

1) clear segmentation throughout the body (segments are added each time they molt until becoming mature),

2) two pairs of legs per segment, and

3) antennae.

Just like Milton, when in danger, millipedes often

curl up in a spiral as a defense strategy to protect the softer, more vulnerable areas with their hard outer shell. While their name suggests they have 1000 legs, millipedes typically have only between 40 and 100 legs, with the record being 750 legs!

Class Malacostraca (Pronounced mal-uh-kos-truh-kuh)

This group contains a lot of different animals, some that live on land and others that live only in the water. The animals in this group include crabs, lobsters, shrimp, crayfish, and even roly-poly bugs. Members of this group have:

1) gills to breathe with,

2) segmented bodies (some species and body region segmentation are more apparent than others),

3) antennae, and

4) three body regions (head, thorax, abdomen).

As mentioned in the story, even the members that

live on land must stay moist to breathe with gills. Many species in this group have pinchers used to capture food and defend their resources from others. Francis Fiddler Crab has one really large claw and one really small claw. Only male fiddler crabs have these enlarged claws, and they wave them in displays toward females.

Class Insecta (Pronounced in-sek-ta)

This group is the largest Class of arthropods. The animals in this group include beetles, bees, butterflies, ants, praying mantis, flies, grasshoppers, and many more. Members of this group have:

1) three main body regions (head, thorax, abdomen),

2) six legs, and

3) wings at some stage of life.

Flying has given this group a tremendous advantage and allowed them to evolve a wide variety of species that inhabit a wide variety of areas across our planet. Insects are the most diverse group on the planet. There are about 1 million described species of insects, and they are found on every continent, even Antarctica.

Class Arachnida (Pronounced ah-rak-ni-duh)

This group contains organisms that look very different from one another, including spiders, scorpions, amblypygids, vinegaroons, solfugids, harvestmen (daddy longlegs), ticks and mites, and others. Regardless of their vast differences, they do have some basic things in common with each other that no other groups in this book share:

1) two body regions, a cephalothorax (fused head and thorax regions) and an abdomen,

2) eight legs, and

3) two extra pairs of appendages not talked about in this book, pedipalps (Savanna's "hands" in

the illustrations) and chelicerae, which are the only pair of appendages in front of the mouth. Chelicerae are often thought of as jaws in arachnids, but spider chelicerae have fangs that allow them to deliver venom to their prey.

Arachnid appendages have evolved into a variety of shapes to perform a variety of functions, including pinchers, walking legs, and even long whip-like legs used to sense their environment.

* * *

By observing and comparing the characteristics that she had with the characteristics specific to each group mentioned above, Savanna was able to determine what Class of arthropods she belonged to. This activity is the same thing that scientists, called taxonomists, do to classify living organisms they collect or see in nature.

About the Author and Illustrator

Author Dustin Wilgers — I am a native of Wichita, Kansas, and have spent the majority of life in the state. I have always been interested in science and nature. I received my B.S. in Biology from Southwestern College, M.S. in Biology from Kansas State University, and Ph.D. in Biology from the University of Nebraska.

It was at Nebraska where I fell in love with spiders. After getting over a mild case of arachnophobia, I became fascinated by these creatures and have been studying their behaviors for over a decade now. I enjoy sharing my passions with others through teaching students of all levels.

I am a faculty member in the Natural Sciences Department at McPherson College, but during my spare time, I hold outreach events across the state focused on conservation and spider biology.

My wife and two kids are my world. Our most cherished memories always seem to include spending time together in nature or at sporting events.

Illustrator Nate Knott — I was born and raised in the Sandhills of Nebraska. I've always loved art and was fortunate to have great teachers along the way. I received my B.A. in Art Education from Doane College. For the last 15 years I've worked mostly with clay, making functional pottery, but have always loved working two dimensionally as well. I think art has the unique ability to help us to slow down and appreciate the world around us. I'm very passionate about learning, and right now I'm teaching high school art, helping others find their passion.